SPACE TELESCOPES

Table of Contents

The James Webb Space Telescope launched in 2021 and is about 1 million miles (1.5 million km) from Earth.

Space **telescopes** are launched into space. They give us views of planets and stars. Scientists observe them from the ground. The telescopes send pictures back to Earth. They help us study things we cannot see from Earth.

telescope an instrument that uses lenses or mirrors to make distant objects appear larger

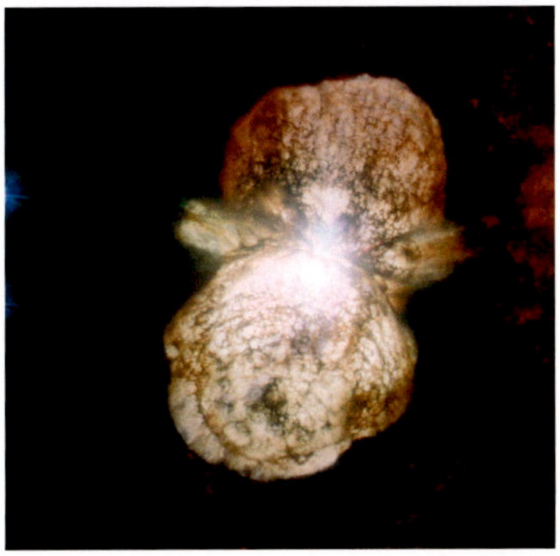

Space telescopes let scientists see a supernova, which is when a star explodes.

Telescopes can help us study Earth from space. Some telescopes are aimed at the Sun. Others are pointed at different planets. Early telescopes detected light that we can see with our eyes. Newer ones measure different types of energy.

Space telescopes allow scientists to see beyond Earth's atmosphere.

More than 90 telescopes have been put into space. Many are no longer active. Rockets launch them out of Earth's **atmosphere**. Most telescopes **orbit** Earth. Others orbit the Sun.

atmosphere the mixture of gases surrounding the Earth

orbit to move in a path around an object

OAO-2 was nicknamed Stargazer.

In 1968, NASA launched the *Orbiting Astronomical Observatory-2 (OAO-2).* It worked for four years. The telescope sent back many pictures. Some showed comets surrounded by clouds of gas. It was the first time scientists saw that!

NASA The National Aeronautics and Space Administration is a U.S. agency that explores space

Explosions on the surface of the Sun are called solar flares.

Some telescopes look at the heat of distant stars. Another kind measures X-rays and **gamma rays**. This energy is found in explosions on the surface of the Sun. Sometimes, these rays keep radio signals on Earth from working.

gamma ray invisible ray that is sent out from harmful radioactive materials

Space telescopes have many parts. Mirrors collect light from stars. Larger mirrors collect more light. Solar panels collect energy from the Sun. Batteries store the energy. This powers the telescope. Scientists use a thruster to change the telescope's direction.

It costs billions of dollars and takes many years to build a space telescope.

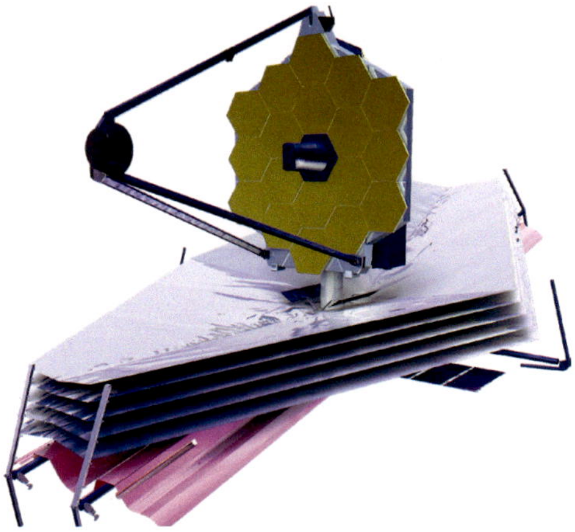

The James Webb Space Telescope has a sunshield the size of a tennis court.

Telescopes have cameras that take pictures. Information travels from the antenna to computers. Special science instruments measure **radiation**. Shields protect the telescope from too much light or heat.

radiation harmful electromagnetic energy particles traveling from an object in space, examples include X-Rays, lasers, and microwaves

The *Hubble Space Telescope* launched in 1990. Since then, it has looked at black holes. It took images of Pluto. It recorded asteroids crashing into Jupiter. *Hubble* orbits Earth 15 times a day. It still sends pictures and information.

Hubble is about the size of a school bus.

Hubble tracks Jupiter's Great Red Spot, a giant storm that is always swirling.

Telescopes give us information about space. We have learned about the temperature of stars. Telescopes have shown the types of gases around different planets. We are learning more about objects in our solar system and beyond!

Spotlight: James Webb Space Telescope

NASA launched the *James Webb Space Telescope* in 2021. It was made to look deep into space. It allows scientists to learn about how planets and stars change over time. The telescope orbits around the Sun. It follows a similar path to Earth. The telescope has taken pictures of faraway galaxies. It has even recorded the changing seasons on Saturn!

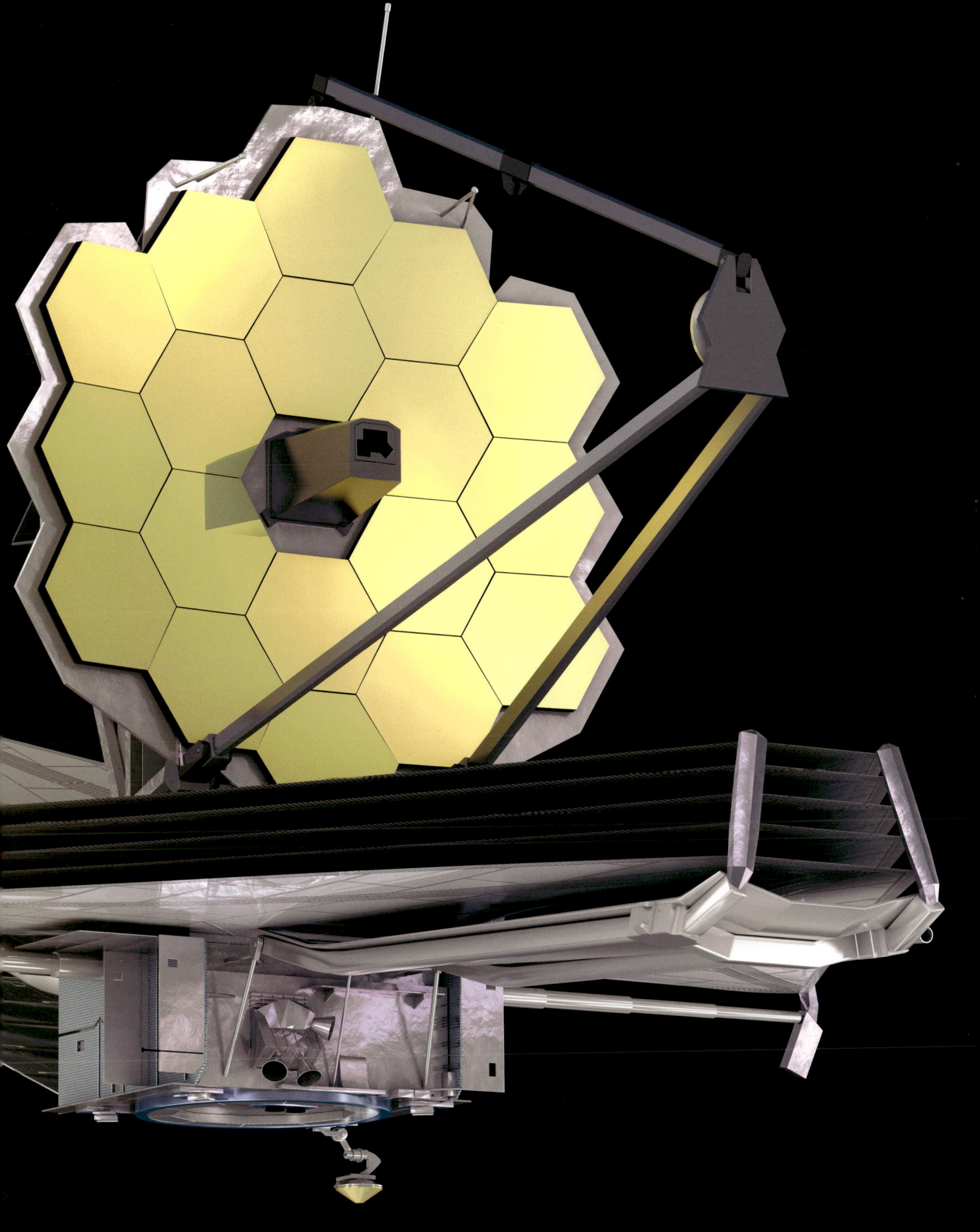

Read More

Bell, Samantha. *Telescope.* Cherry Lake Publishing, 2024.

Peterson, Christy. *Cutting-Edge Hubble Telescope Data.* Minneapolis: Lerner, 2020.

Reeves, Diane Lindsey. *James Webb Space Telescope: A Peek into the First Galaxies*. Lerner Publications, 2024.

Websites

NASA Science Space Place
https://spaceplace.nasa.gov/menu/play/webb/
Learn more about the James Webb Telescope and play games.

YouTube: SciShow Kids - Telescopes in Space! How We Study Space.
https://www.youtube.com/watch?v=XB0Er2eZn54&t=411s
Watch a video about telescopes.

Note: Every effort has been made to ensure that the websites listed above are suitable for children, that they have educational value, and that they contain no inappropriate material. However, because of the nature of the Internet, it is impossible to guarantee that these sites will remain active indefinitely or that their contents will not be altered.

Index